Misfortunes on the Mighty

A Harrowing Voyage Down the Mississippi

Joseph Solomon

The Mississippi River, 2,350 miles of natural beauty, along with dangerous elements. A stretch of the United States that highlights an incredible part of the country. Full of wondrous people, and even more wonderful sites. After two years of planning I was finally taking on the Mississippi River, from source to sea. This would be nothing like my Ohio River trip that I did two years prior. This would be three times as long than my first ever long paddle journey down a river system. I would have to brave the elements and keep my head on a swivel to make sure that I would make it to the end. I promised myself that I wouldn't quit on a bad day, and since every day was a bad day, I wasn't able to quit.

It took me 15 hours to reach Lake Itasca Minnesota from Cincinnati, the source of the Mississippi. On the way there I remember, seeing a black bear in the headwaters, and then an eagle fly overhead. Hopefully they were good omens of the natural wildlife that I would be encountering. I spent the night thinking about all the things that could go wrong on my journey. It was hard to fall asleep. I don't know if it was nerves or excitement. All I knew was that I was about to undertake and immense challenge, one that I was willing and anxiously eager to take.

The next day, I said goodbye to my friends that got me to this point. I launched with three other paddlers, I thought that `ogether we would complete the Mississippi river, but the river had other plans, and it all started when I fell in the water. This was the start of my journey.

Copyright © 2024 Joseph Solomon
All rights reserved.
ISBN- 9798304424356

DEDICATION

I want to dedicate this book to my Mom, Teresa and Dad, Tim. My biggest supporters and my dedicated followers.

Misfortunes on the Mighty

Contents

Misfortunes on the Mighty — i
Acknowledgements — viii
1. Solitary Voyage — 1
2. Treacherous Currents — 7
3. Unexpected Encounters — 9
4. Tranquil un tranquil Respite — 19
5. Unforgiving Storms — 24
6. Reflections on the River — 31
7. The River Tells a Tale — 36
8. Resilience Tested — 40
9. Homeward Bound — 42

Misfortunes on the Mighty

Acknowledgements

Thanks to Leland Walker, Justin Rodriguez, Lauren Phipps for help with proofreading and editing.

Misfortunes on the Mighty

1. Solitary Voyage

I started my trip out with three other guys. We paddled together for a week, but I noticed they were struggling, I was having a hard time myself. I felt my body getting used to the motion of paddling off the bat, but it wasn't just paddling. Parts of the headwaters had dried up, so sometimes I was hiking through muck to get my kayak across a certain part of the river. This wouldn't have been so bad if I didn't have a kayak full of gear, even the rolling portage cart that I brought, broke down on me. A full week of this was a difficult reality that I wasn't expecting to face. Every evening I lay down to rest, but that didn't take away the soreness that I felt every night. I just wanted to be alone. After a week of paddling with these gentlemen, I said goodbye for now. I made my choice, It was a difficult decision, but I needed to rest my body.

At this point, things started getting a little funky. Before crossing Lake Winnibigoshish, we ended up going through a rather strong storm. I am now alone and vulnerable. Everything I thought would be trip ending actually happened during my trip. Sometimes we think of things, and they just manifest themselves.

The next morning when I woke up at the campsite at the fish resort at the end of Lake Winnibigoshish, I decided that I was going to go fishing. So I asked the gentleman that owned the property if he would take me down the street to get some bait, fishing license, and a fishing rig. When I opened my wallet to pay for the stuff, I looked in it and noticed that all of its contents were gone. When the gentleman drove me back to the fishing resort, I tore everything I had apart looking for my debit card, license, and credit cards. While frantically going through my stuff, It clicked in my head where all my shit was. It was at the last campsite where the Mississippi meets the lake.

On this day, it was very windy and there were no fishermen on the lake. I went to the owner of the fish camp and asked him if I could rent a John boat from him.

I knew exactly where my stuff was. It was all the way at that camp that we had that night before, just inside Lake Winnibigoshish, from where the river enters the lake. All of the fisherman were looking at me like I was crazy. I was going out on that lake no matter, what the weather was, I was going to find my stuff. I can't go very far without my license and credit cards.

I started the John boat and went across the lake on my quest to find my wallet contents. As I trudged forward, I was quickly reminded how small I am. The winds were whipping the lake mist in my eyes and the waters were swirling and wild. The waves could've easily capsized me if they wanted to, but I was on a mission. Still though that didn't stop me from praying to make sure I made it across safely.

When I got to the point where the Mississippi meets the Winnibigoshish Lake, I remembered, the night before when we camped there, one of my friends had chopped down a small tree. This was so that all of our tents would fit on this rock wall. As soon as I saw the down tree, I knew I was exactly where I needed to be.

I dropped anchor jumped in the water and swam to the site where we had camped the night before. See, the story behind this is we were in such a cool looking spot, that I got into the lake the night before to take a picture of our campsite. I threw my wallet towards my tent, and apparently everything inside of it went flying out.

I found all of my wallet under a bush, then I proceeded to get back in the John boat and went to the Fish Camp. Let me tell ya, it was it a ride to cross that lake inside of that John boat. The water from the props kept coming into the back of the boat. I was racing the waves as they cross the lake, just as I was.

After about an hour and a half of crossing, I safely arrived back at camp. What a relief it was to have everything back, and what a relief it was to have navigated that lake in such horrid conditions.

There was no fishing boats on the lake that day, nor was there any kayaks or canoes. As I walked back into the lobby, there were fishermen standing everywhere and they were looking at me as if I was crazy.

One gentleman asked: "Are you the guy that just went out there in the lake in that John boat?"

I replied to him: "Yes. I had to get my wallet and it was the only way I could get it".

Another guy from the crowd yelled out, "Dude! You're my hero".

I don't know if I'm such a hero, but it was pretty wild being on the lake that day.

During my first week of paddling alone. I went through some really hard times. Things were happening that would set the stage for some really disastrous happenings along my journey. Things that would make anyone quit.

Right off the bat somebody called me a train wreck which eventually became my new river name. This is a story of train wreck moments as I traveled down the Mississippi River.

The first train wreck moment was losing everything in my wallet, and it certainly would not be my last. From the get go, I was determined to paddle the Mississippi River, in solitude, packed with determination, and I thought: "I will never quit on a bad day".

Furthermore I had the mindset that bad things were coming almost every day. It was very difficult to get past these feelings. I can't name names because I'm terrible with remembering them, but I will tell you stories about going down the Mississippi river, stories where I went through some really hard trials and tribulations to get to my destination: The Gulf.

2. Treacherous Currents

No more than 2 minutes into the journey, there's a low-lying bridge where you can do the limbo and go under it or you can portage around it. I decided to do the limbo, but I lifted my head up too soon. As I was going under the bridge I hit the main support, put my arm up to brace myself, and there was just enough current to twist my boat around and dump me in into the river.

I don't know what was worse my ego or my determination. It was an embarrassing happening and I don't like the fact that it happened so quick.

After flipping back and cussing up a storm I pumped all the water out of my kayak, I gathered all my belongings scattered in the waters. Including my electronics that were wet. I realized that this was probably not too bad a thing to have happen to me. It let me know I was not at all prepared for a trip down the most iconic river in the United States.

Mother nature, is a place of solitude and healing. It's very humbling that I was even here. Nine years prior I had had a near death experience. Now I'm out here paddling a whole river system. There are a lot of dangers out here. To be able to do it is about 98% mental. The irony is: I have a mental health disability and I was paddling to raise awareness for people with mental health conditions, whom in the same boat as I.

From Lake Itasca all the way to the Gulf, the river winds back and forth and it's very unpredictable. It could be a metaphor for the unpredictability of my mental health conditions.

From Lake Itasca, to St. Louis there wasn't much of a current. The waters were at record lows. There was nothing to assist me to get from the lake all the way to St. Louis. It was slack waters. With no current, I had to depend on my own muscles to get me across this river. From St. Louis south to the Gulf, it's a free flowing river. At times there were numerous treacherous wing dams. Which are known to make their own currents that can suck your kayak, underwater and drown you. I had to do everything in my power to make sure I avoided them at all cost, but sometimes you have to share the river with large barges. Their large wakes were also a hazard to my little kayak. This didn't stop me from continuing on with my mission.

3. Unexpected Encounters

My first week alone, I stopped for lunch one day and had some trail mix that had some really crunchy pieces. As I was chewing, I felt a tooth in my mouth crack in half. I thought to myself "Are you freaking kidding me?"

This tooth had been in my mouth for an entire 41 years, it was a baby tooth. To be able to have kept it in my mouth for 41 years is kind of a miracle. As soon as the tooth cracked in half I knew I was in trouble. My gum swelled up and I was in a lot of pain.

I stopped at River angels house and I asked them if they knew of any dentist within the area. I called every single dental place where they lived and nobody could take me in. I thought to myself this could be trip ending. If I don't get this tooth fixed, it's gonna make it difficult for me to finish my paddle.

Then I got a call from somebody, a random woman, saying: "Hey the next town you're going to, there's a dentist there. I lied to them and said I was a patient and you were my little brother and that it's time to go see the dentist."

Shoot, I didn't even know I had a sister, but today I was gonna be her little brother, so that I could get my tooth worked on.

They told me it was gonna cost $600 to take the tooth out. At that point, in my head, this was trip ending. When I went up to the counter to pay my $600, It was only $20. I started crying because I felt like they had just given me a huge donation.

I met with another River angel who asked me to stay with him for a few days while I got this tooth taken care of. It takes a few days of healing before I can actually get back and paddle again. This gentleman took care of me for four days. I really appreciated that because of him and his generosity I was able to shower and wash my clothes.

Life has a funny way to force you to relax, even when you want to keep paddling. Either way, I'm thankful to have spent time with this river Angel.

After my tooth had healed and the sutures were gone, it was time to get back on the river. I had made it on the news shortly after I had the surgery on my tooth.

When I was paddling one day through the windy river, I came around the Oxbow on the northern part of the Mississippi. I saw a family of people standing on the riverbank.

"This might be a time to stop to talk to somebody" I thought to myself.

There was a gentleman in the back of the crowd of people standing on the riverbank that yelled out "Hey Joe, how's your tooth?"

So random that I'm in the middle of nowhere, and all of a sudden somebody says something about my tooth. They just saw me on the news that morning, and the tooth was mentioned. I instantly pulled my kayak over and stopped to talk to these folks.

They were having a graduation party for one of the teenagers that had just graduated high school. They were so kind they fed me and in return, I told them stories of my journey. Once I had my fill, I thanked them and I moved along my way.

I proceeded to go down the river, and met a few more river angels along my route. One River Angel reached out to me and said that there was electricity to charge up and a shower, If I chose to take a shower. Another paddler said it was a good place to charge up.

So I proceeded to do my laundry at this house and charge all my batteries up. I was going to continue on for that day. The laundry machine stopped, and I took my clothes out and put them into the dryer.

What I did not know, is the laundry machine did not do its full cycle. I had four pairs of clothes packed with me and would change them every so many days. As I traveled further down River, this is when things started getting a little crazy.

I had met up with two Paddlers in a canoe, and we decided that we were going to paddle with each other. When you're in a kayak all day, alone, it's always nice to see other people. I paddled with these two gentlemen for four days and on the third day we were paddling, and I noticed that my bottom side had hurt really bad.

I chalked it up to sitting all day every day. As I paddled with these guys, and I had to paddle harder because two men in a canoe, move faster than one man and a kayak.

I noticed my bottom side was really on fire and had no clue why. We stopped at River angels house and one of the guys got a hold of some medication to put on it.

We thought I had a fungal infection, because when you sit all day, you sweat and it can cause fungal infections. So I sprayed some spray on my ass and it started burning worse. I couldn't even sit down.

We pushed off in the morning, and made it across Lake Peppin. There was an island at the end of the lake. This is where we were going to camp for the night. We stopped on that island. I pitched my tent and I was just in this horrifying pain. I got into my tent took my pants off that I had just changed three days before took a picture of my bottom side. Sure enough it was a massive scab, then I looked into my pants that I had just taken off, only to find a Tide Tod inside of my pants. I thought back, and that tide pod came from the River angels washing machine.

Somehow the washing machine failed to do the full cycle and the Tide pod was stuck in my pants, I had this horrible chemical burn on the bottom side of my body. It hurt so bad. I called my friend Justin to tell him about it. I was ready to quit. I was over it but talking to him made me feel better even though it was a crappy situation. I knew all I needed was a little time to heal.

Those two gentlemen left the next morning. Alone again I thought to myself except this time it wasn't from exhaustion, but instead, a chemical burn. The only way to explain a chemical burn, is burn, literally. It was like the universe was trying to make sure that I did this journey on my own. I stayed and I laid there in pain for four days until my bottom side had healed up. After resting and scrolling on Facebook for four days it was time that I moved on from that island.

I then paddled along for a few more days, which were sweltering hot. It was difficult because I had to mind my water intake, to make sure I had enough before I stopped into town again. At least now, though I didn't have to worry about chemical burns on my butt, that was a plus I guess. That was one of the few train wreck moments I wasn't expecting. I don't think any other Paddler has had that type of experience at least not to my knowledge.

After a few days of paddling, I came up to a lock and dam. I called the lock Master to see if I could get through. This was about 6 PM in the evening. There was traffic on the upside of the lock where I was, and apparently there was a lot of river traffic on the low side of the lock. The Lock Master told me that it was gonna be about six hours for me to get through at that point.

I felt very defeated. I asked the lock Master if I could camp on top of the long wall.

His response was: "Yes, just don't build a fire."

I respected his wishes, and pitched on a slope and proceeded to sleep for the night. I kept rolling down the slope not only that but at night I would get eaten alive by thousands, of mosquitoes. Between them, biting me and flying around my ears and the slope that I was on it was hard for me to get any sleep till eventually Mr. Sandman got the best to me.

The next morning I woke up early to get through the lock because I knew that there would be less traffic that time in the morning. The Master told me it should be clear when I got there.

I grabbed some of my gear and proceeded down the riprap, which are giant boulders, where my kayak was tied up floating in the water. As I was going down the riprap, I noticed a large square concrete block, which some barges used to tie up if they can't get in or out of the lock. I accidentally brushed up against this concrete block, looked down and noticed that there was a swarm of wasps around my legs.

I knew at that moment I was done, these wasps are gonna sting me and I just have to accept it. Thank the Lord above, it got down to about 64° the night before. The wasps were still a little sluggish, because it was early in the morning and it was cool at night.

The second time I looked down at my legs. I noticed there was a whole nest of wasps circling my legs, then 1 came in and stung me. Then they all started stinging, and at that point, I ran up those giant boulders as fast as I could screaming and swatting to get the wasps away from me.

I got back to where I camped for the night, turned around to look where the wasp were at, and saw that there was a whole swarm of them protecting their nest inside of that concrete block.

They were blocking my way back to my kayak. So I had to go down the Rock wall, further down away from the wasp swarm, swam in the river with my gear and pack it into my kayak.

The process took me about an hour and a half. Once finally back in my kayak, I called the Lock Master, and they let me into the lock. I asked the Master if they had any Benadryl that I could send down to me, his response was they don't have stuff like that in their emergency kit.

I then immediately reached out on the Internet to any river angels that were south of where I was located. There was a river angel in the next town, which was about 4 or 5 miles from the lock. He reached out to me and said that he could take me into town to get some Benadryl and other supplies. I met this gentleman at the boat launch, in the town that he lived in. He brought a tarp to cover my kayak while he took me into town.

Now there are a few things that I take with me when I go into town. So they don't get stolen. I was only gone for 15 minutes, and when we got back to my kayak, I noticed something was missing.

Somebody had stolen a bunch of stuff out of a box that I had strapped to the back of the kayak. The river angel was very kind and asked what had been stolen from me. My response was they stole all of my pot and anything that has to do with pot like my pipe, my edibles, and my flower.

He said: "Well, you can't call the cops on that one".

At this point I realized that I had had a horrible day. I just got stung by a bunch of wasp, and I was falling apart, and my medical marijuana I use to soothe my horrible PTSD, was now gone. He suggested I come home with him and rest for the day. I accepted his offer and went home with him.

4. Tranquil un tranquil Respite

After the tumultuous events of the past few days, I welcomed the opportunity to rest and recharge along the banks of the Mississippi. I had been paddling tirelessly, pushing myself to the brink of exhaustion, and my body ached with every stroke of the paddle. As I guided my kayak towards a secluded cove, I could feel the tension slowly leaving my muscles, replaced by a sense of relief and tranquility.

The cove was a hidden gem, its still waters reflecting the surrounding trees and the clear, azure sky above. I carefully maneuvered my kayak to the shore, securing it to a sturdy branch before stepping onto the soft, sandy bank. I took a deep breath, savoring the fresh, earthy scent that permeated the air, and felt the weight of my journey begin to lift from my shoulders.

Eager to fully immerse myself in the serenity of the moment, I set about making camp. Not paying attention that I was in a low lying area. It was a beautiful camp spot.

I was unaware of an impending storm that was coming right at me.

As the sun began to dip below the horizon, I settled myself in my tent fixed on the mesmerizing dance of the river. The gentle lapping of the water against the shore were the only sounds that interrupted the peaceful silence. I felt a sense of profound gratitude for this moment of respite, a chance to recharge and reflect on the challenges I had faced thus far.

My mind wandered back to the treacherous currents and unexpected encounters that had tested my resolve. I recalled the moments of sheer terror, when I had feared for my life, and the instances of pure serendipity that had saved me from certain disaster. Yet, through it all, I had persevered, driven by an unwavering determination to complete my journey down the mighty Mississippi.

As the night deepened, my body was finally surrendering to the exhaustion that had been building. I drifted off to sleep, lulled by the soothing sounds of the river and the knowledge that I had earned this well-deserved respite. Tomorrow, I would continue my voyage, but for now, I could rest, my spirit renewed and my resolve strengthened by the tranquility of this fleeting moment.

What came next was far from tranquil. I had no cell phone service to watch the weather or see what was coming at me. The River angel that had dropped me off that afternoon realized that I only camped right on the other side of the lock where he had left me. He sent me a text message saying there was a massive storm coming, right at me and somehow I was able to receive a picture from him with the storm cell and what it looked like on the radar.

At that moment, I knew I was really in for it because I was not protected. I was on top of a low lying sandbar and the storm just looked like it was wretched, then I got another text from him. Be careful Joe this storm is very strong, and then it hit. It was a torrent of a storm filled with lightning, heavy downpours, and heavy winds. My tent was flopping in the wind all over the place. The thing about staying in a tent is you really can't see outside of it so I lifted the vestibule back and looked outside and noticed that the sandbar I was on, had standing water on it. This was not a good situation. I immediately started removing things from my tent and placing them on a tree, which was on higher ground. My tent had about 4 inches of water on the inside of it because this low lying sandbar was disappearing as the storm raged. I didn't get much sleep that night, but knew that I had to paddle the next day no matter what. If I had a bad night, I still had to get forward movement the next day.

Misfortunes on the Mighty

5. Unforgiving Storms

The next day, The calm, placid waters that had carried me so steadily the day before had been replaced by a turbulent, churning flow, the surface whipped into a frenzy by the strengthening winds.

Reluctantly, I began to pack up my campsite, knowing that I needed to make haste if I hoped to outrun the disappearing sandbar. As I secured my gear and double-checked the lashings on my kayak, I couldn't shake the uneasy feeling that had settled in the pit of my stomach. The river, once a source of solace and adventure, now seemed to hold a dark, unpredictable power that filled me with a sense of dread.

With a deep breath, I pushed my kayak into the water and climbed in, my paddle dipping into the choppy current. The going was slow and arduous, the strong winds and swelling waves were making it a constant battle to maintain my course. I strained against the elements, my muscles aching with the effort, but I refused to give in to the temptation to seek shelter and wait out the storm.

As the day wore on, the conditions only worsened. The skies opened up, unleashing a torrential downpour that drenched me to the bone and reduced my visibility to a mere few yards. The river, once a familiar friend, had become a treacherous foe, tossing my small craft about like a toy in the hands of a capricious giant. Navigating the treacherous waters became an exercise in pure survival, as I fought to keep my kayak upright and on course. The waves crashed against the hull, sending spray cascading over me, and the wind howled with a ferocity that seemed to mock my efforts. I gritted my teeth, my focus laser-sharp as I battled the elements, determined to reach the safety of the shore.

Time seemed to slow to a crawl, each stroke of the paddle a herculean effort. My arms burned with the strain, and my lungs ached with the effort of drawing breath in the face of the relentless wind and rain. But still, I pressed on, my resolve hardening with each passing moment.

As the storm raged on, I found myself navigating through a veritable gauntlet of obstacles - fallen trees, submerged debris, and treacherous eddies that threatened to pull me under. I maneuvered my kayak with a deft touch, my years of experience on the water serving me well as I deftly avoided the hazards that loomed before me.

Yet, even as I fought to maintain control, I couldn't help but feel a growing sense of despair. The river, once my ally, had become my adversary, and I wondered if I would have the strength to overcome the relentless onslaught of the elements. The thought of being swept away, lost to the

churning waters, filled me with a primal fear that
threatened to overwhelm my senses.

But I refused to give in to the darkness. I drew upon the deep well of resilience that had sustained me through countless challenges, and with a renewed determination, I pressed on. Each stroke of the paddle was a testament to the unwavering spirit, a defiant refusal to surrender to the unforgiving forces of nature.

As the hours ticked by, the storm showed no signs of abating, and my body began to betray the toll that the ordeal was taking. My muscles ached with fatigue, and my hands, blistered and raw from the relentless grip on the paddle, threatened to give out. But still, I pushed forward, my mind focused on the distant shore that seemed to taunt me with its elusive promise of safety.

Finally, after what felt like an eternity, I caught sight of a small inlet, a sheltered cove that offered a respite from the raging storm. With a surge of renewed energy, I steered my kayak towards the haven, my heart pounding with a mixture of relief and trepidation.

As I beached my craft and stumbled onto the shore, I collapsed to my knees, my body trembling with the sheer force of the ordeal I had endured. The storm raged on, the wind and rain battering the land around me, but in that moment, I felt a sense of profound gratitude for the simple act of being alive.

Slowly, I dragged myself to my feet, my gaze sweeping across the small, sheltered cove. It was a haven, a place of respite in the midst of the unforgiving chaos that had threatened to consume me. With a weary sigh, I set about securing my kayak and making camp, my movements slow and deliberate as I sought to regain my strength.

I huddled in the relative safety of my makeshift shelter, my mind replaying the harrowing events of the day. I marveled at the sheer power of the river, a force of nature that had tested the limits of my skill and determination. And in that moment, I felt a newfound respect for the Mississippi. A respect born of the knowledge that the river was not to be trifled with, but rather navigated with a reverence and caution that only experience could instill.

Yet, even as I contemplated the perils I had faced, I couldn't help but feel a sense of pride in my own resilience. I had stared down the storm. I had fought against the relentless onslaught of the elements. I had emerged victorious by the slimmest of margins. It was a testament to the strength of the human spirit, a reminder that even in the face of the most daunting challenges, the will to survive could triumph.

As the night wore on and the storm began to subside, I found myself drawn to the edge of the cove, my gaze fixed on the now-calmer waters of the river. The once-raging torrent had been tamed, its fury spent, and I marveled at the transformation, at the way the river could shift from a terrifying force of nature to a serene, placid expanse.

In that moment, I had a determination to continue my journey down the Mississippi, to face whatever challenges lay ahead, with the same unwavering spirit that had carried me through the unforgiving storm. I knew that the river would continue to test me, to push me to the limits of my endurance, but I was ready. My resolve hardened by the trials I had already overcome.

With a deep breath, I turned my attention to the task of preparing for the next leg of my journey, my mind already racing with the possibilities that lay ahead. The river, once a source of trepidation, had become a canvas upon which I would paint the next chapter of a remarkable odyssey. A testament to the power of the human spirit to conquer even the most daunting of obstacles.

6. Reflections on the River

As my journey down the mighty Mississippi continued, I found myself drawn deeper into the rhythm of the river. The steady cadence of my paddle dipping into the water, the gentle lapping of the waves against my kayak and the ever-changing landscape that unfolded before me became a soothing symphony that calmed mind and nourished my soul.

In the quiet moments, I had ample time to reflect on the experiences that had led me to this point. The challenges I had faced, the unexpected joys I had discovered, and the profound sense of connection I felt with the river and the land it traversed all coalesced into a tapestry of personal growth and self-discovery.

The solitude of the river provided a rare opportunity for introspection, and I found myself delving deeper into the questions that had long lingered in the back of my mind. What was the true purpose of this journey? What had I hoped to find or to learn? As I paddled on, the answers began to emerge, slowly and organically, like the unfurling of a flower in the morning light.

I recalled the initial trepidation I had felt, the doubts that had crept in as I prepared to embark on this adventure. But now, those fears seemed distant, replaced by a sense of unwavering determination and a profound appreciation for the lessons the river was teaching me.

Even with the treacherous currents I had navigated. The unexpected encounters. I had learned to read the river's moods, to anticipate its shifts, and to respond with a level of agility and grace that I had never before possessed.

As I drifted along, my gaze would often drift to the shoreline, where I would observe the ebb and flow of life. The small towns and bustling cities that dotted the riverbanks, the wildlife that thrived in the riparian habitats, and the ever-changing patterns of the landscape. All served as a constant reminder of the interconnectedness of all things.

In these moments, I was nothing but a small speck on the vast canvas of the river, yet I was inextricably linked to the greater whole. The Mississippi had become a teacher, a guide, and a companion on this journey, and I knew that the lessons I was learning would stay with me long after I had reached my final destination.

As the sun began to dip below the horizon, casting a warm glow over the water, I paused to take in the beauty of the moment. Sunsets and sunrises along the Mississippi were all so breathtaking. I knew that the challenges that lay ahead were still formidable, but I also knew that I was better prepared to face them than I had ever been before. With a sense of purpose and a heart filled with gratitude, I dipped my paddle into the water and continued my journey down the river, my mind and spirit forever changed by the reflections on the river.

7. The River Tells a Tale

The Mississippi River flowed steadily, its waters glistening in the morning sun. I dipped my paddle into the current, propelling the kayak forward with each stroke. As I navigated the winding waterway, I couldn't help but feel a sense of awe and reverence for the mighty river that had become my companion on this journey.

Each bend in the river held a new story, a new adventure waiting to unfold. I had learned to listen to the river, to read its subtle cues and understand its moods. The river was a living, breathing entity, and it had much to teach those who were willing to listen.

As I glided along, My mind drifted back to the tales I had heard from the locals along the way. The Mississippi had seen it all – from the early Native American tribes who had relied on its bounty for centuries, to the European explorers who had charted its course, to the modern-day adventurers and thrill-seekers who sought to conquer its challenges.

The river had witnessed the rise and fall of empires, the ebb and flow of human civilization. It had seen the birth of towns and the abandonment of once-thriving communities. The stories it held were as vast and complex as the river itself.

I thought about the people I had met, the stories they had shared. I remembered the weathered fisherman who had regaled him with tales of the legendary catfish that lurked in the river's depths, their whiskers as thick as a man's arm. I recalled the riverboat captain who had spoken of the many paddlers he had met through his years on this mighty river.

Then there were the stories of the river's own resilience – how it had endured floods and droughts, how it had adapted and changed with the passing of time. I had seen firsthand how the river could be both a source of life and a force of destruction, a provider and a taker.

As I paddled on, I felt a deep connection to the river. The river was a storyteller, and I was honored to be its scribe, recording the tales that unfolded before me.

The water lapped against the sides of my kayak, and I could almost hear the river whispering its secrets to me. I leaned in, listening intently, as if the river itself were imparting a sacred knowledge that only a true devotee could understand.

In that moment, I felt a sense of belonging on this plain. I was no longer just a solitary traveler on the river; I was a part of something larger, a participant in the grand narrative that the Mississippi had been weaving for centuries.

I knew that the river had a lot of stories to tell, and I felt I didn't have enough time to hear them all. I was limited by what I was able to soak up along my journey.

For me, the Mississippi was more than just a river – it was a living and breathing. A repository of human history and a testament to the enduring power of nature. As I paddled on I realized the river was much than just a waterway to navigate, it is a part of a much larger system that feeds water to the entire US. The greatest waterway in America. I used it as a vessel to get my message out to the public about Mental Health Awareness. That was the passion and beautiful part of my epic Journey down the Mighty Mississippi River.

Misfortunes on the Mighty

8. Resilience Tested

From a broken tooth to a tide pod incident that gave me a chemical burn on my ass, I still pushed on. I was tested time and time again by the unpredictable weather and being exposed to the elements for long periods of time.

The river kept trying to test my patience, and I was not going to succumb to her relentless overpowering of my mission. When your out there day in and day out grinding to get to where you need to be, there is no planning that can go along with the unpredictable nature of such a massive beast. I would honestly say that the River is a beast that demands your respect. If you slip and veer away from that respect, you're gonna mess around and find out she doesn't care.

From 80mph winds with no rain, which battered my tent one night, to the Bear that decided to sniff around my camp in the headwaters, I can't think of a time that I didn't have the rivers full attention. She is demanding of your full attention.

I was tested physically, emotionally, mentally, and the hardest test was spiritually. This Journey, single handedly changed my life in un explainable ways. I love this river and Its majorly unpredictable course.

I spent a lot of time, money, and hard work to do this trip only to make it to Vicksburg Mississippi. I was unable to continue my journey because there was a hurricane that came through, hurricane Ida. She was a nasty storm. A friend of mine took me inland to get away from the hurricane, the next morning after the hurricane struck, I knew that my journey had to come to an end.

There was no way I was going to paddle into a disaster zone looking for electricity, water, and possibly shelter while everybody else was looking for that themselves. Out of all the things I experienced that happened to me along this journey I did not give up. But when it came down to it, mother nature had a different idea in mind for me. She really tested my resilience hard, and I'm not going to compete with a hurricane.

9. Homeward Bound

The suns golden glow over the mighty Mississippi as my, battered but resilient boat was ready to go home. After the harrowing trials and tribulations I had endured over the past months, the prospect of finally reaching the safety of home filled me with a profound sense of relief and gratitude.

My friend Matt, from Cincinnati, was in Jackson Mississippi with his semi and flatbed, anxiously waiting to take me home.

Even the tranquil respites I had found along the way had been tinged with a sense of unease, as I knew that the river's calm facade could quickly give way to the unforgiving fury of its storms. Yet, through it all, I had persevered, drawing strength from the very waters that had sought to claim me.

Now, as the distant lights of my hometown came into view, the river, I realized, had not only tested my mettle but had also imparted upon me a deeper understanding of the world and my place within it. The tales it had told, the lessons it had taught, would forever be etched into my memory, shaping the man I had become.

The Mississippi had been both my adversary, and my companion, a force to be reckoned with and a source of solace in equal measure. In many ways, it had become a part of me, and the thought of leaving it behind, even temporarily, was a difficult one to bear.

Yet, I knew that my journey was not yet complete. There were still miles to be traveled, challenges to be faced, and stories to be told. The river, in all its might and mystery, would continue to beckon me, and I was determined to heed its call, time and time again.

As I stepped onto the familiar ground of my hometown, I was greeted by the warm embrace of my loved ones, their faces alight with joy and relief. In that moment, I realized that the true measure of my resilience was not found in the battles I had won, but in the connections I had forged - the bonds that had sustained me through the darkest of times and the most treacherous of waters.

The mississippi prepared me onto my next great adventure.

ABOUT THE AUTHOR

    Joseph Solomon is an outdoorsman and adventurer. He lives on rivers. He is an advocate for mental health awareness. He loves to serve his community. A former nurse who has served communities all around the United States, including the Navajo nation, and continues his long journeys for mental health awareness.

Made in the USA
Columbia, SC
12 January 2025

50738032R00033